JAKU 618

10 Ways Internet Improves Your Life

Contents

1

Introduction

Humanity has come a long way since the stone ages. From sticks and stones to now sending spaceships and satellites to outer space. Humans have changed since then centuries ago. Even looking back to the past forty years or so a lot has changed since the invention of the internet. Since the invention of the internet on January 1, 1983 the standard of living has become better and more sufficient. It has brought culture much closer to each other by making it easier and faster to connect to people from far away. With better access made knowledge and learning more accessible with the vast amount of information there is on various topics you could think of. To say that the invention of the internet is life changing is an understatement. Some would even go far as to say its a cultural artifact in of itself.

The Internet is mostly known as the interconnection of computer networks. It's a massive network working with other networks. Connected by millions of computers worldwide globally forming a large interconnected net in which any computer can communicate with each other so long both are connected to the internet. Sometimes it's called the Network or Net for short. This network is so important for our daily lives as it offers so many services and resources we use for our daily living. It's critical in this day and age to utilize the internet to our advantage so we can move up in life.As it's a tool for us to link to other people from far distances and create our own mark on the world.

But how do you use the internet more effectively and use the world wide web to your advantage? We spend a lot of our valuable time on the internet throughout our day so in order to spend your precious time being as productive and efficient as we can this collection will help you to be a more effective user of the web. Most people probably know what the internet is but in summary the world wide web is for the exact purpose of communication across large or even small ranges to share information from any place in the world and have access to said information. It is freely available to the vast majority in our world that has access to wifi. The main fields I'll go over in this book are education and research, community, money management, business, entertainment, travel, health, and conveniences the internet provides.In This book I'll give you some tips and advice on how to improve your life in each field using the wonderful tool we call the internet.

2

Education

Knowledge is a major factor in our big blue world and is the key to our growth as mankind continues on living. It's a tool for people of all ages to learn throughout their lifetime growing older. Everyone can use the internet to learn new or relearn stuff. Many can seek education and training classes available in online courses. Most devices nowadays are connected through the Internet. The Internet has broadened the availability of educational content on any topic. Internet search engines help people quickly find the relevant study material in multiple formats for example such as videos or research documents. This solves the issue of going to the library to read several books just to find relevant information they were looking for. Gathering intel is on the tips of your fingertips with just one google search you can seek what it is you were looking to learn about and the info is usually on the front page or for a more detailed answer a scroll down.I have listed down five strong reasons why education was impacted probably the strongest the most in the years the internet was created. Here are four reasons why having internet access has impacted education as a whole.

Easy Accessibility To Knowledge

As stated before the internet's unique access to knowledge is among its many significant educational contributions. The day when students were restricted to wisdom in their classrooms or libraries has long since passed. The internet makes a multitude of information available. Students can now explore their interests and learn more about a wide range of subjects outside the confines of the traditional curriculum, thanks to the availability of scholarly papers, eBooks, and online courses.

Learning Together

Student learning together more closely is an educational approach that encourages students to actively engage with their fellow students and resources to collectively construct knowledge and problem solve. With this approach, students work together in a group or team, sharing ideas, gaining new perspectives, and responsibilities to achieve common learning goals. Through the provision of several digital tools and platforms that enable seamless communication, information sharing, and collaborative projects, the internet has significantly aided collaborative learning.

Students can communicate and work together regardless of where they physically are through online discussion forums, video conferencing, document sharing, and real-time collaboration tools. By allowing students to learn from each other's viewpoints, give and receive criticism, and build upon one another ideas, collaborative learning fosters critical thinking, communication, teamwork, and problem-solving abilities. It encourages active participation and engagement in the learning process while creating a supportive and engaging learning environment that prepares students for issues they will face in the real world.

Learning Becoming More Personalized

The internet has made education also more individualized and adaptable to different learning styles. Utilizing data analytics, online learning platforms, and adaptive technologies personalize student learning opportunities depending on their individual skills, shortcomings, and preferred learning styles. This personalized approach allows students to learn independently, delve deeper into areas of interest, and receive targeted support. Internet-based learning also offers flexibility, allowing learners to access educational content anytime and anywhere, accommodating diverse schedules and preferences.

Worldwide Multicultural Exchange

Worldwide multicultural exchange describes the synthesis of many cultural perspectives and cross-cultural knowledge sharing. Since the advent of the internet, academic institutions and students worldwide have been able to communicate, work together, and share knowledge. Thanks to this exchange of ideas and cultural awareness, the educational process is enhanced, tolerance is promoted, and students are better prepared to become global citizens in a world that is growing more interconnected.

Thanks to the internet, students may communicate across cultures, learn about other customs and beliefs, and develop a more comprehensive awareness of world concerns and gain worldly knowledge. The internet is crucial in altering education and educating students to flourish in a diverse yet connected global society by encouraging a worldwide viewpoint and cultural interchange.

3

Research

When we talk about research capabilities, we mean that students and researchers are better at being able to use the internet to access and acquire information, carry out in-depth research, and investigate a variety of themes and issues. The availability of massive databases, digital libraries, academic publications, and other readily accessible online resources has revolutionized the research process thanks to the internet.

A researcher can explore various viewpoints, collect data, and more quickly and effectively analyze findings because of easy access to information. The internet also allows researchers to collaborate with specialists and academics from different fields, promoting knowledge sharing and interdisciplinary study. With improved research capabilities, scholars and researchers can learn more about their areas of specialization, keep up with the most recent advancements, and advance knowledge in their respective domains.

In which the internet has broadened access to information it makes it easier for the average person to develop answers to our world around us. This makes the internet an excellent tool for researching day to day problems. Numerous topics ranging from science to politics. The

Internet is an amazing research resource that offers access to various materials on almost any subject. Articles, books, and many other information sources are more accessible than ever because of the growth of web databases and libraries. To add on the internet is pivotal for a researcher's collaboration and communication the particular reason for this circumstance is that it allows them to exchange thoughts Some people would bring up the internet as not always a trustworthy source depending where you get the information from so it's good to check where you get your intel from. It is critical to assess the material thoroughly and keep the source reliability in mind.

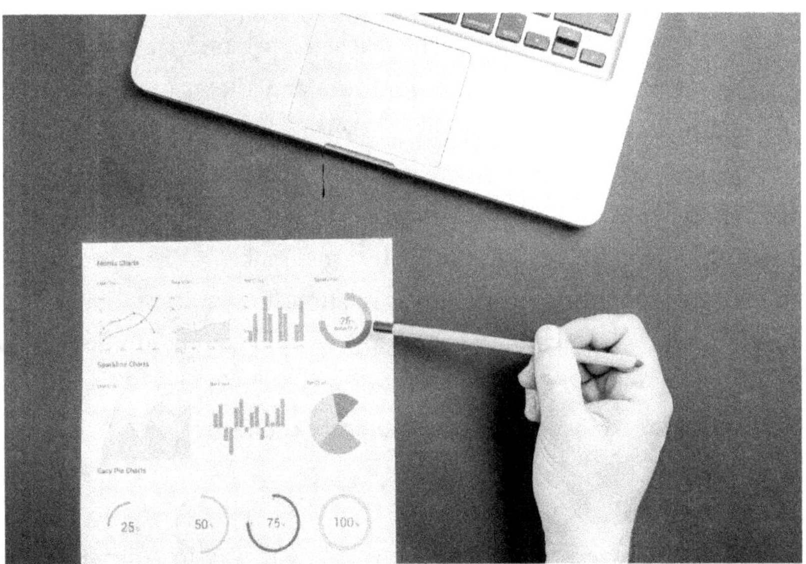

4

Community Building

As the internet expands the reach of a person by a huge margin. It helps people to communicate and allows us to share the information we have gained with a large number of people. One person can be on the other side of the planet and could be able to speak with one another in real time.Be it through text, email, or speech through a phone call. Access to the Internet makes it easier to keep in touch with loved ones and make new relationships. Your reach is pretty much unlimited which means you can broaden your community you have around you and take that as an advantage. Email and Social being the two major factors of how a community can form.

Email

Email is a more cost-effective and efficient way to communicate with others across the web using the internet.By using email instead of traditional mail you can save more space and materials. Due to the way email is so easy to use and free, many people use it to check their email on their computers and carry their phones, which allows them to always have access to their email inboxes frequently. It's also free and available across different platforms, pretty much anyone can have an email address.

For you to build a community a crucial feature you'll need is email which is important for communication because it lets people send information in a letter and send mail to someone. You can send updates to family,friends, and/or other associates.Use it to sign up for accounts of many websites. As well for security reasons to ensure the right user has received your message. Along with text you can send documents, images, and videos to another user. As it is set up you are allowed time to think on the content you want to mail to the other person about and respond back more thoroughly to the reply.

Social Media

Social networking can be used to interact with other users or to find people with similar interests to yourself. The purpose of a social network is to share, learn, interact and market a product.Social networking sites have connected people around the world. Social networking is an essential part of the Internet.

With the help of the Internet, people have got the ability to form social groups where they can share information, thoughts and ideas about any-thing. The social networking platform is the largest source of content, covering everything from informative content to entertainment. The best thing is that people don't have to pay anything to use these services. This helps businesses to develop their community and promote their products. Social media is the one big factor I must talk about that has impacted on how to build a community of like minded individuals that share the same interest as you.

Many social media accounts along the lines of facebook, myspace, discord, youtube, instagram, and formerly twitter now called X are platforms where many people with commonality can be brought together and converse among the many topics the individuals really fond over while also branching off to make new relationships.Finding a close-knit community can help make you feel more valued and accepted over the negative thoughts that could foster in your head that you aren't. Even for pre-existing relationships, social media makes it easy to keep those relations connected as family members separate and part ways.The other ways social media can have an impact on us is how our self-esteem and mental health can improve. As well how influence behavior can have an effect on shaping personal growth.

5

Money Management

As much as the internet is useful for the exact purpose of finding info, it's also great to store money away with banking websites, apps, or any other tools that help us in transferring,managing, and budget. Access to the internet is helpful for an individual's economic growth and finance stability. Online money management allows you to access your account without wasting cash. It also helps keep track of your spending history or patterns. Online banking has many features such as overdraft protection alerts and paperless billing. It's also extremely safe so long as you follow some precautions along the lines of long easy to remember passwords or questions only you and close ones would know. You can also budget your money with the help of these online banking apps to help track and organize your expenses, offer insights regarding spending habits, and set aside a portion of your monthly income to a savings account. These apps can also read your bank details, show where you can make better savings, and help us understand the truth behind the financial data. They're also free for the most part financial advisers that are there to help break down your money to decide how and where to spend it better.

6

Business

The internet is a booming success for anyone looking to start and operate a business.Due to the internet there's been a rise in increase in entrepreneurship and self-employment. This is the particular reason that many resources are now widely available to anyone so anyone can start a business. It made it easier to file taxes and get your finances together in order. Even operating a business is much easier due to the increase of automation from the internet reducing the need for less busy work. Organizations and businesses can also use email to communicate about upcoming projects and tasks within the organization with decent security. This can be an effective and safe way for team members to communicate, especially when some members of the team are working at home as a remote job or not in the office. From the boss to the employees all benefited by the access of the world wide web. By selling products from the various online markets it's a new service and opportunity creating new jobs. For someone who is unemployed as well can benefit from the internet as many businesses are looking for employment if they are able to work. Find a job training or look up the position online to see if the business is hiring. You can create, update, and improve your resume by applying to many jobs over the course of your lifespan. Access to the Internet can help you improve your skills and find your next job. The internet also gives small businesses a platform to make them more visible to the average person and more likely to receive customers online. The size of the company doesn't matter or if the workers are a team of introverts the web is a powerful tool to connect as many people as possible.In addition to reaching a broader scope of potential consumers, having a digital presence can as well help small businesses build brand awareness and credibility. Three major internet features that have improved businesses are marketing, online reviews, and automation.

Marketing

One of the biggest advantages created by the internet is the ability to access and market to large audiences. Content marketing is essentially free, and businesses can create content that their audience desires, as a means of driving qualified traffic to the business. Search traffic is a powerful method of introducing a business to an audience. If a business does not want to spend the manual hours chasing down organic search traffic, they can purchase top locations in the search results, and essentially buy segments of that specific audience. The audience is using specific search terms relevant to the business, so that the traffic is extremely qualified. The rate is typically built on bidding platforms with more competitive keywords garnering higher values. Additionally, a business can purchase banner advertisements through high traffic websites, purchase viewers on social media platforms and utilize influencers and public relation tactics to drive website traffic.

Review Sites

Review websites and review options on business listings have had a major impact on a business growth as consumers can research through positive and negative reviews before making a purchase decision. This works to your advantage as the business with great reviews most of the time offers good customer service. Compareto the disadvantage of a business with bad reviews with below average score usually having either subpar products or bad customer service. Review ratings tend to have the option for a business owner to respond and it reflects on their ability to engage and communicate with the customer.

Automation

Automation is the ability to automate tasks that usually require a workforce that ultimately saves businesses a lot of money and makes their operations simpler and more efficient. Everything from basic accounting practices to customer service has some automation abilities depending on the business model and requirements. A small business can automate receipt tracking and basic accounting through software services that connect to accounts and track and classify spending and income.

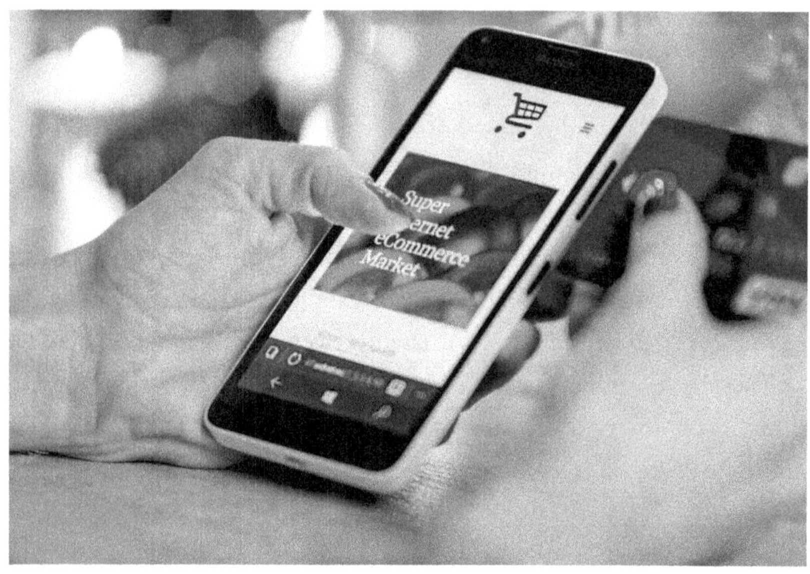
Super
ernet
eCommerce
Market

7

Entertainment

Entertainment is all around us more so than ever due to how widespread the internet is. It's relaxing to just kick back and enjoy the moment listening to some chill tunes or watching an exciting movie. The internet allowed people to entertain others or be entertained by constant involvement such as video games or youtube videos.The online media is an ever growing vast variety of subjects. You might want to create content on such platforms to gain a following whether it's through art or broadcasting yourself to the world to hear your voice. Many people would support and even donate you money. It's been made possible because the internet is designed to make it easier to target and grow an audience world wide.

Entertainment in Social Media

As far as entertainment is concerned, social media is known for its widespread content. It could be hilarious or informative.This actively demonstrates that the typical type of venue is outdated for targeting just one city or area when an entertainer can instead advertise themselves or their performance on social media and live broadcast. Changing the way entertainment is viewed, received, and consumed by a mass audience. Also thanks to the internet you can now watch replays of events which usually wasn't the case back many years ago. With the advent of the internet, we are now ushered into the age of streaming. With tons of quality streaming services such as Amazon Prime Video, Netflix, and Hulu, streaming shows, and movies, our time of leisure has become fun. Furthermore, we can stream the content of our choice wherever and whenever we would like. You can browse and watch the trailers of upcoming movies and make an informed decision about whether you want to catch the movie in the cinema or not.

8

Travel

Travelers from all over the world use the internet to research and explore the destination before you set off on your trip. It'll help you on how to organize and set specific preferences and needs you want. Among the most notable uses of this is the option of shopping and making travel reservations online. You can explore different websites and platforms dedicated to buying a flight, booking hotels, car rentals, search tourist activities, and travel packages, allowing you to compare prices and find deals to suit your type of budgets and travel preferences.

One of the best ways to uncover authentic places and find local recommendations is through the internet, as the web offers a wealth of information provided by locals willing to share recommendations about their hometowns.You can get recommendations through apps such as TripAdvisor, Yelp, or Google Reviews, where you may find opinions and feedback from other travelers about restaurants, hotels, tourist attractions, and areas of interest, although the best of all is the rating system, which will help you pinpoint the places most recommended by the locals.Other ways to get local recommendations are by reading blogs or travel sites of bloggers and frequent travelers, who share their experiences online with in-depth recommendations, which helps spot local events, festivals, and holidays.

23

Booking online has many advantages and benefits, including flexibility when changing, canceling, and access to exclusive discounts. The Internet has made it a lot easier for people to book tickets for buses, trains, and domestic or international flights directly using their devices from anywhere. People can also book an uber or taxi by choosing their current location to a set destination where they'll be picked up or dropped at a specified location. No more waiting in long queues for their turn to book tickets at the ticket counter.

9

Health

The idea that people are responsible for actively managing and maintaining their personal health and well-being was greatly boosted by the internet. Your health is very important and with the internet you have access to the world wide web which can help you keep track of important health goals, find information on diseases and treatments, and access your medical information such as medicare or healthcare providers. Searching for new recipes to meet weight-loss goals, tracking weight loss, and learning weight loss activities. Encourage people to take responsibility for their own health with wellness programs and incentives.Current users can monitor their health and do daily activities from online exercises courses. Internet users can also monitor their diet and exercise online with the help of virtual networks and communities, but they can also access internet tools that catalog their personal health information to help them identify and manage their risk for depression, diabetes, or the flu.The Internet has streamline health care while improving the overall world health with better healthcare providers.

Through the internet you can help manage and catalog your personal health information and keep track of your process. With the help of a virtual community it's easy to have others to encourage your healthy lifestyle. Actively managing and maintaining their personal health and well-being - known in the medical community as self-management - is getting a boost through the internet. Employers already encourage their employees to take responsibility for their own health with wellness programs and incentives.

10

Conveniences of the Internet

There's so many other conveniences the internet provides that I didn't even go over like world wide news, internet shopping, smart home technology. All of which can be used in your everyday life. If you can spend money in a store it's very likely to have an online website where you can purchase your goods from home or at work.Besides, people can order a wide variety of products at home using the Internet and devices. It can range from grocery products to ready to eat, fashionable clothes to medicines. Most items can be ordered at home and received directly at the door. You can be up to date with the latest news from trusted news networks online free of charge in the form of podcasts or videos. With smart home technology, internet-connected devices known today as smart home technology can reduce many everyday home maintenance tasks. These devices use your internet connection to help you live a more efficient lifestyle while in your home. Smart devices along the lines of refrigerators that organize grocery lists, door locks that alert you of disturbances and medication dispensers that ping your pharmacist for the exact purpose of a refill can give older folks all the tools and independence they need to age in place comfortably. For examples of smart home devices that make your home management a little easier.

- Home security – You can keep an eye on your house with home security devices like smart cameras and doorbells. They typically have a companion mobile app that you can download to see live streams and receive notifications when there's activity while you're away or an alarm when you're asleep.
- Health monitoring sensors – Wearable health monitoring devices are a non-invasive and cost-effective way to keep tabs on your wellness. They can track active heart rate.
- Automatic medicine dispenser – Automatic medication dispensers keep you on schedule with your meds. The dispenser sends alerts to you or your family if you've missed a dosage and shares detailed reports with your doctor.
- Smart light switches – Have control over your lights with set timers or voice commands from your mobile app or smart home device hub. Smart lights can also make it look like someone's home when you're away, reducing the chance for burglars. They can reduce energy costs and set the mood for morning,bedtime, and special events.
- Key finders – Trying to find keys are the most aggravating things to do as keys are the easiest to lose track of. To locate them more easily, you can use a Tile or other key finder device and app to ring an alarm to help you find it. Some also show you where it is on a map.
- Smart thermostat – Smart thermostats learn your preferences and help you cut heating and cooling costs. You can program it yourself or let it automatically optimize for comfort and cost savings. Many have a companion mobile app and voice command features.
- Smart home hub – Living on your own has become a lot more simpler and more convenient with a central smart home hub to

control everything for you. You can sync up your thermostat, entertainment system, light switches and among other devices and control it all with voice commands.

- Automatic stove control – Automatic turn-off devices for when you aren't sure you turn off the stove. They typically come with a timer, motion sensor and automatic shut-off feature.
- Wi-Fi refrigerators – A smart fridge can store your grocery list and share it with participating grocery stores nearby.

11

Creating a better world

Even with so much already known about our world it's still in its early stages as far as the world wide web has to offer and it'll continue to expand and diversify into many different forms along the lines of self driving vehicles, automated mining, and real time machine language translation glasses. Throughout the years huge changes will occur which will allow advances in day-to-day areas like along the lines of health care, safety and human services, will continue to have a significant measurable improvement in many lives, reduction in bad outcomes and incidences. Advances in opportunities in education, community and creative work will continue as time goes on. While it's difficult to predict exactly what the future of the internet will look like because new technology is evolving so quickly. There is no doubt that the newest iteration of the web will transform virtually every part of our economy and society as we know it. I believe it'll be for the better as life has gotten much better for the average human being in the past 20 years than it has ever for humanity in the last hundreds of years.

Empowering people with a sense of agency to do the right things for thousands to even billions of people around the world.The internet has exposed people to multiple viewpoints and beliefs creating more aware individuals about our world. As well as being introduced to compassionate people that are all fighting for justice and bringing peace.The internet enables us to come together with people all across the globe. No matter the race, gender, or belief we'll come together for a common cause. This actively demonstrates that we can highlight and fight against injustices, better understand people's needs, and connect with people we may have never otherwise met. We can therefore conclude that the internet has made society better. Over the past decade, people's optimism to help solve the world's most pressing issues has increased over time. It's all thanks to the internet providing us with the knowledge and resources to handle mostly anything that we are likely to come across in this world. As much as there is evil in the world there's also an equal amount of good to counter balance it. The internet does show us the good and bad of humanity as a species but as we continue to grow more aware of our surroundings the better we can make the world for ourselves and future generations.

12

Conclusion

As we've reached the end of this book I hope you will see how much of a world phenomenon the internet is and will be in humanity. Going from cavemen with sticks and clubs to global technology. Humanity has come a far way since then and will continue to push onwards as time moves on. With these tips and advice are to help those who didn't grow up with this tech or to freshen why it's so important to those who have taken it for granted.To sum up everything that has been stated so far the world wide web is life changing tech it has profoundly impacted our world in countless ways. It has affected how we communicate, learn, work, manage finances, accessible access to entertainment and conveniences, and still continues to evolve to shape us and our big blue world. I hope anyone who has read this book gets out from reading this that the positives of our world have gotten only better and may continue to do so for many years to come. By utilizing this amazing technology we've gained people are able to make huge substantial progress in our lives. The ways we can help improve ourselves and others by using the resources that are just in the palm of our hands. Only we can shape that future for ourselves.

I hope you enjoy reading this book as much as I loved writing about it. If you found this book to your liking please leave a positive review on how this book has helped your life even a little bit.